YOUR KNOWLEDGE HAS VALUE

- We will publish your bachelor's and
  master's thesis, essays and papers

- Your own eBook and book -
  sold worldwide in all relevant shops

- Earn money with each sale

Upload your text at www.GRIN.com
and publish for free

Optimizing Analog Communication Systems. A Simulink-Based Approach for Bandwidth, Frequency, and Waveform Analysis

Bandar Hezam

Bibliographic information published by the German National Library:

The German National Library lists this publication in the National Bibliography; detailed bibliographic data are available on the Internet at http://dnb.dnb.de.

ISBN: 9783346980090
This book is also available as an ebook.

© GRIN Publishing GmbH
Trappentreustraße 1
80339 München

Print and binding: Books on Demand GmbH, Norderstedt, Germany
Printed on acid-free paper from responsible sources.

The present work has been carefully prepared. Nevertheless, authors and publishers do not incur liability for the correctness of information, notes, links and advice as well as any printing errors.

GRIN web shop: https://www.grin.com/document/1426542

Acknowledgments

First and foremost, I would like to thank my lecturer for Communication Engineering Principles Dr. TIANG for her guidance and advices that assisted me to finish this lab report as well as her explanation of the concepts needed for this lab report.

I also would like to thank her for clarifying and explaining me some examples that related to the topic of my lab report. Besides, special thanks also to my senior colleagues at A.P.U, especially group of Engineering Department for sharing the literature and invaluable assistance.

I would also like to convey thanks to the Management of A.P.U. for providing the laboratory facilities.

Finally, an honourable mention goes to my parents and my elder brother in particular who support me to study abroad. Without help of these who mentioned above, I would have faced many problems while doing this lab report.

Table of Contents

List of figures

Introduction

Communication can be defined as the process of transmitting information and common sympathetic from one world to another or from one person to another. It is one of the most vital processes of a human's life. It can be completed by using remote communication which aids to join them to each other or to transfer the information from one world to another. The communication system involves a lot of separate transportations (communications) networks, substations, diffusion systems, receiver stations, and data terminal apparatus typically accomplished of interoperation and interconnection to generate an integrated system of the total. Most of the communication systems reduction into one of three types: the efficiency of the power, the effective of the cost or the wide band's efficiency. Power efficiency shows the system's capability to correctly transmit data or information at the lowest applied power level, when the efficiency of bandwidth of a modulation system shows and describes the capability to supply data in a specific bandwidth. (Hanzo, Webb and Keller, 2000)

Two popular fundamentals in each communication system are the sender and the receiver. The sender set the communication. In a school, the sender is the person who sends any information in the form of analog and digital communication. The receiver is the individual to whom the message is sent. In addition to that, communication system should have some limited apparatuses whether the system is digital or analog communication system. Apparatuses contain an input transducer, channel, transmitter, receiver and output transducer.

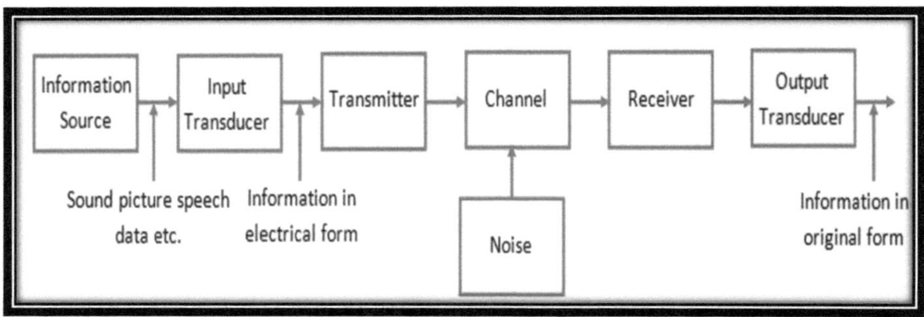

Figure 1: block diagram of a Communication System.

(Electronics Post, 2015)

In this lab, the suitable software used is Simulink. It gives the capability of simulating and control an analog communication system. It describes the characteristics of bandwidth, frequency and waveform of the information transmitted through baseband, modulation cases and construction case.

Introduction to MATLAB & Simulink

MATLAB is a language of technic and computing the information as graphical signals and mathematical codes. It involves a matrix and mathematics numbers to represents the codes which has been created to connect between the imaginary equation and systemically signal. In this software, thousands of the companies developed based on the technical methods used to compute their information.

Simulink, a supplementary product to MATLAB, it enables quick structure of simulated prototypes to explore design ideas at any level of detail with negligible effort. It is also considered a block diagram field for multinomial reproduction and modulation any design. It provides imitation, automatic code production, and constant test and confirmation of fixed systems. It used to model, to simulate, and to analyze dynamical schemes. It provides linear and nonlinear schemes, modeled in incessant time or a cross of the two systems. Engineers in everywhere use Simulink to get their ideas off the pounded, involving reducing the fuel productions, emerging safety-serious autopilot software, and conniving wireless and network systems. In Simulink, it is very straightforward to signify and then simulate a measured or mathematical model on behalf of a physical scheme. Replicas are signified graphically in Simulink as diagrams called block diagrams. A huge array of blocks is available to the user in supported libraries for representing many marvels and models in a range of arrangements. (MATLAB & SIMULINK, 2010)

Objective

This experiment has to be done correctly which to test, examine and generate Amplitude Modulation-Double Side Band with Full Carrier (AM-DSFC) signal, and demodulate the modulated signal to regenerate the original message signal by using Simulink.

Introduction to amplitude modulation

The modulation is the process of which the signal can be computed or transferred from one place to another within medium called carrier in a high frequency. This signal can be audio, picture or video. Demodulation is opposite the process of modulation with getting the carrier message and decodes it back into information. (Webb and Hanzo, 1994)

Modulation includes two waveforms: a modulating signal that signify the message, and a carrier wave that outfits the specific application. A modulator methodically changes the carrier wave in communication with the differences of modulating signal.

Analogue modulation may be divided into amplitude modulation (AM) and angle modulation. Amplitude modulation is the procedure of amplitude modulation entails fluctuating the highest amplitude of sinusoidal carrier wave in proportion to the immediate amplitude of modulation signal, on other meaning it's essentially when the amplitude of high frequency carrier wave is changed according to the strength of the signal. It can be characterized as AM with double Side Band with Full Carrier (AM-DSFC), double side-band suppressed carrier (DSB/SC), vestigial side-band (VSB), single side-band suppressed carrier (SSB/SC).

While the amplitude modulation is one of the simplest and easiest forms of signal modulation to implement and display, it is not the most effective and not the most efficient in terms of band efficiency and power custom.

Experiment 1 focuses on the amplitude modulation, and experiment 2 focuses on the frequency modulation

Va

Vc

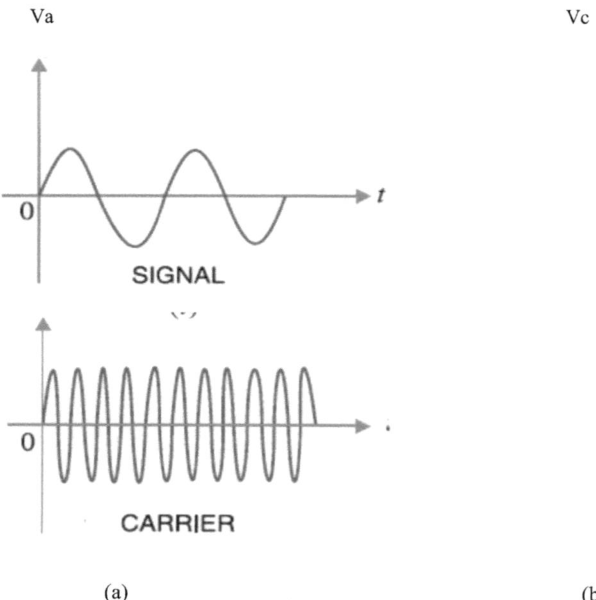

(a)   (b)

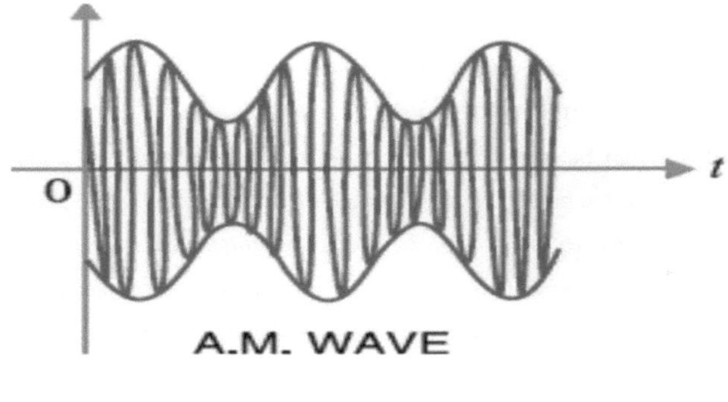

(c)

Figure 2 amplitude modulation.

Modulation Index (m) and frequency spectrum

Typically, the modulation index (m) of a signal fluctuates as the modulating signal intensity fluctuates. And its value is the ratio of the amplitude of audio signal to the amplitude of carrier wave.

$$m = \frac{V_a}{V_c}$$

Assuming the carrier signal is $V_c = V_c \sin\omega_c t$ and the modulating signal is $V_a = V_a \sin\omega_a t$
The amplitude of modulated signal is,

$$A = V_c + V_a$$

$$= V_c + V_a \sin\omega_m t = V_c \left[1 + \frac{V_a}{V_c} \sin\omega_a t\right] \text{ where:}$$

$m = \frac{V_a}{V_c}$ is the modulation index depth.

$$= V_c \left[1 + m \sin\omega_a t\right]$$

Of modulation, $0 \leq m \leq 1$

- The instantaneous voltage of the resulting amplitude modulated wave is,

$$V = A \sin\theta = A \sin\omega_c t$$

$$= V_c \left[1 + m \sin\omega_a t\right] \sin\omega_c t$$

$$= V_c \sin\omega_c t + V_c m \sin\omega_a t \sin\omega_c t$$

$$= V_c \sin\omega_c t + m \frac{V_c}{2}\left[\cos(\omega_c - \omega_m) t - \cos(\omega_c + \omega_m) t\right], \text{ is the equation of an}$$

- Amplitude modulated wave contains three terms:

1- $V_c \sin\omega_c t$, is the voltage of the carrier wave of frequency fc

2- $m\frac{V_c}{2}[\cos(\omega_c - \omega_m)t$, is the lower sideband (LSB)

3- $m\frac{V_c}{2}\cos(\omega_c + \omega_m)t]$, is the upper sideband (USB)

Figure 3 shows the frequency spectrum of the AM wave. It is noticed that the spectrum contains three components. The first component is carrier, and the other two components are called side frequencies.

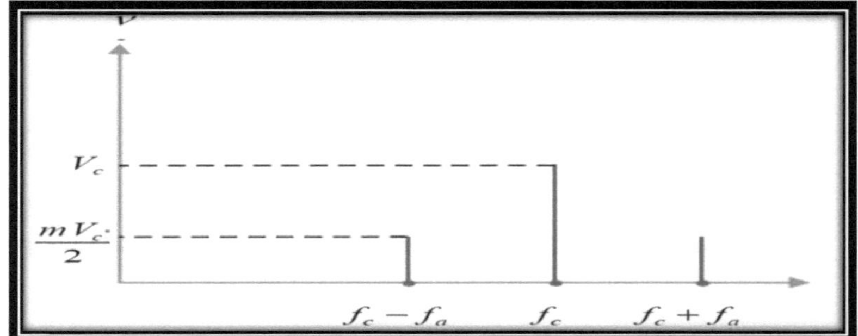

Figure 3 frequency spectrum.

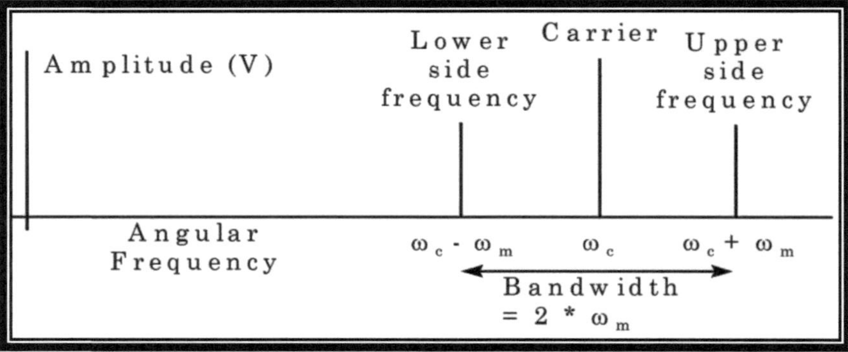

Figure 4: spectrum frequency

The spectrum of these signals is shown. This is labeled as the signal in the frequency domain, as opposed to the signal in the time domain. In this case the audio signal is made up of a single frequency.

In this example the angular frequencies (expressed in Radians/sec, or kRad/sec, or Mrad/sec) are show. In most cases however the frequency is shown (expressed in Hz, or kHz, or MHz).

If the audio signal is made up of a range of frequencies from f1 to f2 (as is normally the case) rather than a single frequency the output signal will be a band of frequencies, contained in

- the upper side band (USB), inverted and
- The lower side band (LSB), erect.

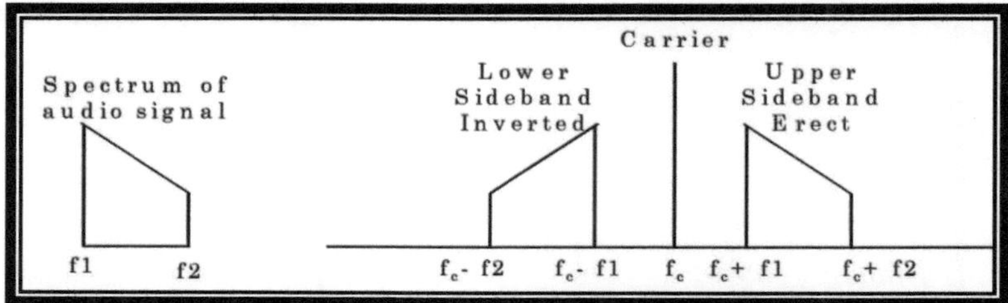

Figure 5: spectrum sideband 2

A broadcast AM station in the Medium Wave band is usually allocated a frequency slot 9 kHz wide. This means that the carriers of stations in this band are spaced 9 kHz apart.

The maximum amplitude in an AM signal is $V_c + V_m$. The minimum amplitude is $V_c - V_m$.

Power in an AM waveform

Assume that the AM signal is dissipated in a load of R Ω. The total power dissipated will be the sum of the powers in all of the components of the signal.

The power in the carrier will be

$$P_c = \textbf{Fehler!} \text{ Watts}$$

The power in each of the frequencies is

$$P_s = \textbf{Fehler!} = \textbf{Fehler! Fehler!} = \textbf{Fehler!} \, P_c$$

The total power is given by the following formula:

$$P_t = P_c + P_s + P_s = P_c + 2\,P_s = P_c\,(1 + 2\textbf{Fehler!}) = P_c\,(1 + \textbf{Fehler!})$$

Watts

The fraction of the power in the carrier is

$$\textbf{Fehler!} = \textbf{Fehler!}$$

Procedures according to using Simulink

Double Sideband-Large Carrier Transmitter (DSB-LC)

The main objective is to build a Simulink model for the conventional AM or the Double Sideband-Large Carrier (DSB_LC) communication system as shown in Figure 4.

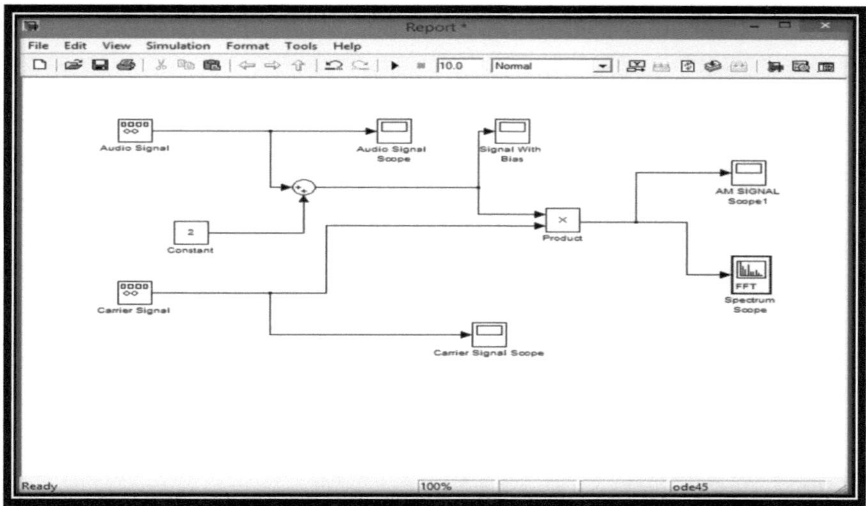

Figure 6: Double Sideband Large Carrier transmitter.

In figure 6 shows the block diagrams used in order to create a double Sideband-Large carrier (DSB_LC) transmitter. The Signal Generator block was used as input data source and the signal parameters were set as amplitude of unity voltage (1 V), the frequency of (10 HZ) and the type of the input signal is sine wave. For the carrier signal generator, the same procedure was repeated the only different is the parameters was set as (2 V) as amplitude and (100 HZ) for the frequency. Constant block has been pulled and the parameter was set as 2. For the sum and product blocks the time was set as (0.001). The scope was dragged from the simulink in order to view the waveforms. Two discrete scopes for the audio signal, carrier and Am signal.

Procedures

By using (MATLAB) and after typing the word of Simulink in the comand window then followed by the following steps:

1. The new model was opened from the menu File followed by (New) and then (Model).

2. (Simulink Toolbox) was selected then (Sources) and then (Signal Generator).

3. (Simulink Toolbox) also was selected then (Sources Carrier) and then (signal generator).

4. (Simulink Toolbox) was selected then (commonly used block) was selected then (Constant).

5. (Simulink Toolbox) was selected then (commonly used block) and then (product).

6. (Simulink Toolbox) was selected followed by (commonly used block) then (sum).

7. (Simulink Toolbox) and the (Sink) was selected then (Scope).

8. (Simulation) was choice followed by (Configuration parameters)

9. Finally, the simulation was run

Results

1. Observing the waveforms and the frequency spectrum

As it can be seen the waveforms and the frequency spectrum for the modulating signal using MATLAB Simulink is shown in the Figure 7 bellow and the observation of the audio signal, signal with bias, carrier signal. Am signal and the power spectrum scope.

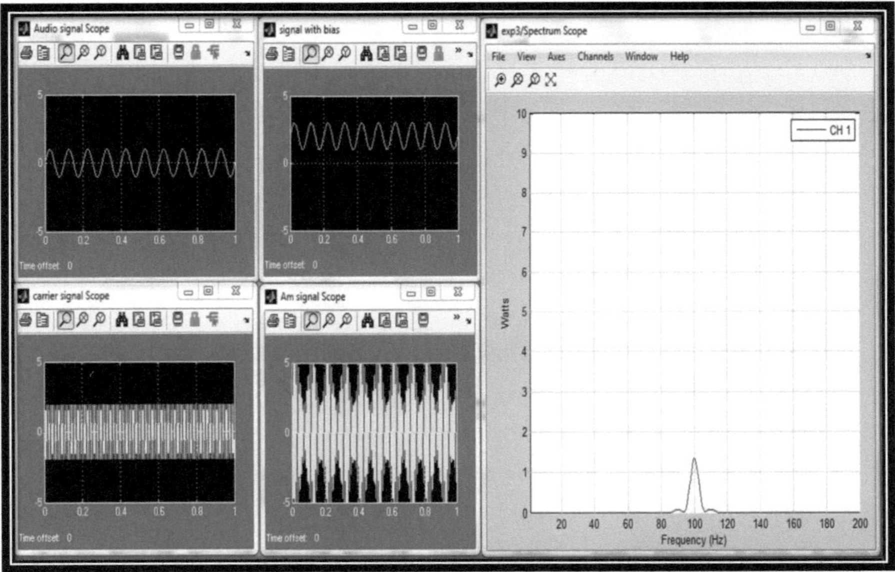

Figure 7 waveforms and the frequency spectrum.

As it can be seen cleerly in the graph above, the spectral output of the spectrum with the input and the output generated signals. Therfore, there are two side components in spectrum shown. First one, components at fc + fm. Second is − (fc + fm) along with a central impulse.

As it can be clearly seen in Figure 8 the waveforms and the frequency spectrum for the modulating signal using MATLAB Simulink, the waveform is display based on lab sheet instruction that given. Moreover, it had been inserting the right parameters in the Simulink and it can be get the following waveform.

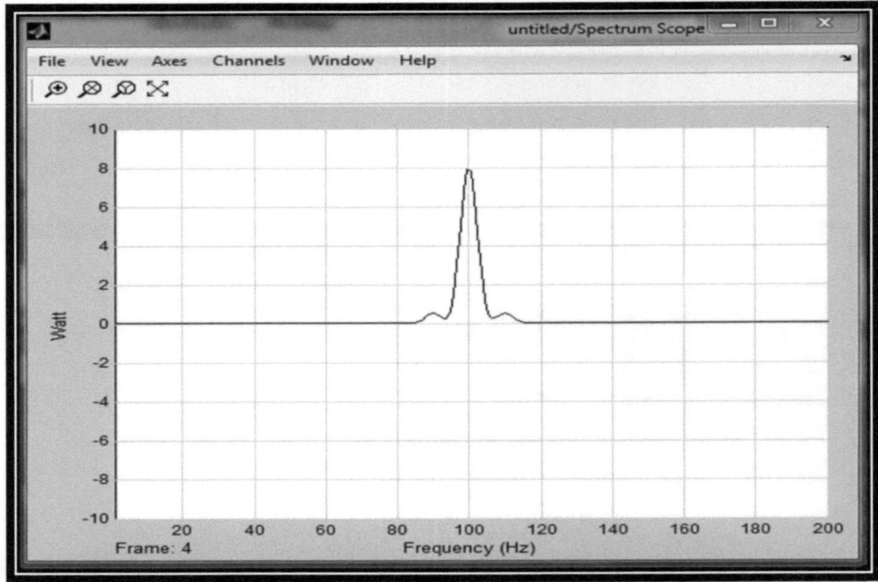

Figure 8 Frequency Spectrum

As it can be seen in Figure () the frequency spectrum of the amplitude modulation. It displays 3 signal which are carrier signal, an upper sideband USB from fc (fc+fm) and a lower side band LSB from (fc-fm) to fc. Adding to that, the signal shows the maximum carrier power at 100Hz is 8 Watts. The upper side band is at 110 Hz and the lower side band is at 90Hz.

2. Calculate the modulation index (m) of the AM signal. Record the amplitude of the carrier, upper sideband and the lower sideband.

Basically, to calculate the modulation index (m) of the AM signal. It can be use this formula

m= (Vmax-Vmin) / (Vmax+Vmin)

Vmax = 9 V and   Vmin = 3 V

Now, we can substitute in the equation above to get the modulation index (m)

m = (9-3) / (9+3)

$$= \frac{1}{2}$$

= 0.5

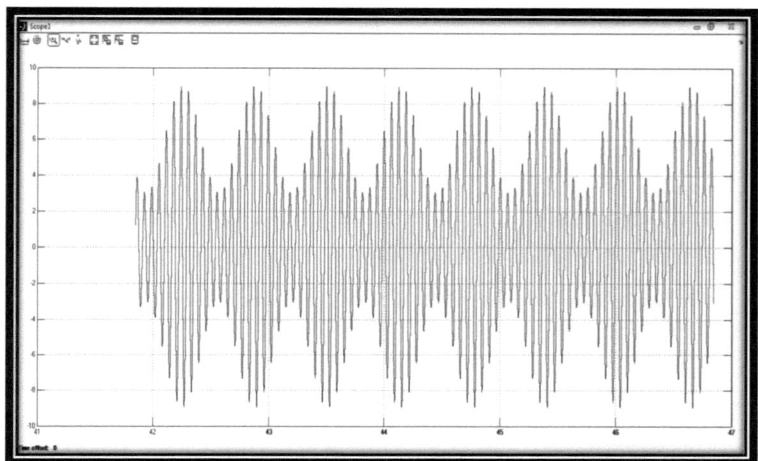

Figure 9 AM signal

3. Change the audio signal and carrier signal parameters to obtain the modulation index of 0.4 and 0.8

To figure out the modulation index to be equal to 0.4 it can be set the parameters as shown below to get the same value.

The parameters for the audio signal and the parameters for the carrier signal respectively are Va = 1, Vc=2 .5.

By using this formula:

$$m = \frac{V_a}{V_c}$$

$$= \frac{2}{5} \quad = 0.4$$

Furthermore, the parameters of the audio signal and the parameters of the carrier signal were changed and it had been figure out a modulation index of 0.4 (by this formula $m = \frac{V_a}{V_c}$ we can find the right values when m =0.4) where Va=1 and Vc=2.5.

To figure out the modulation index to be equal to 0.8 we will have to set the parameters as shown below: (end)

The parameters for the audio signal Va=1.6

The parameters for the carrier signal Vc=2

The formula was used

$$m = \frac{V_a}{V_c}$$

$$= \frac{1.6}{2} = 0.8$$

It can be seen again that when the parameters of the audio signal and the parameters of the carrier signal were changed to obtain a modulation index of 0.8(this formula was used m = $\frac{V_a}{V_c}$ to find the right values when m =0.8) where Va= 1.6 and Vc =2:

4. Calculate the power carried by the carrier and the sidebands

Carrier power can be determined by using the following formula of carrier power:

$$Pc = \frac{1}{2} Vc^2$$

$$= 0.5 \, (2)^2$$

$$= 2 \text{ W}$$

The power of the upper sideband, P_{usb}= The power of the lower sideband, P_{lsb}

Sidebands power can be calculated by using the following formula

$$P_{usb} = P_{lsb} = \frac{m^2 Pc}{4}$$

Hence the modulation index from the experiement is equal to 0.5(m=0.5)

Sidebands Power

$$P(usb) = P(lsb) = \frac{m^2 \, Pc}{4}$$

$$= \frac{(0.5)^2 * 2}{4}$$

$$= 0.125 \text{ W}$$

5. Complete the Simulink model of the AM communication (DSBFC) system by adding receiver model. Capture the input and output waveforms and compare them

A receiver is an electronic device which is used to receive the signals which has been transmitted from the transmitter side, which acts like an encoder and decoder. Usually with the aid of an antenna, propagates an electromagnetic signal such as radio, television, or other telecommunications. The basic receiver system for detecting the input audio signal is represented in Figure 8.

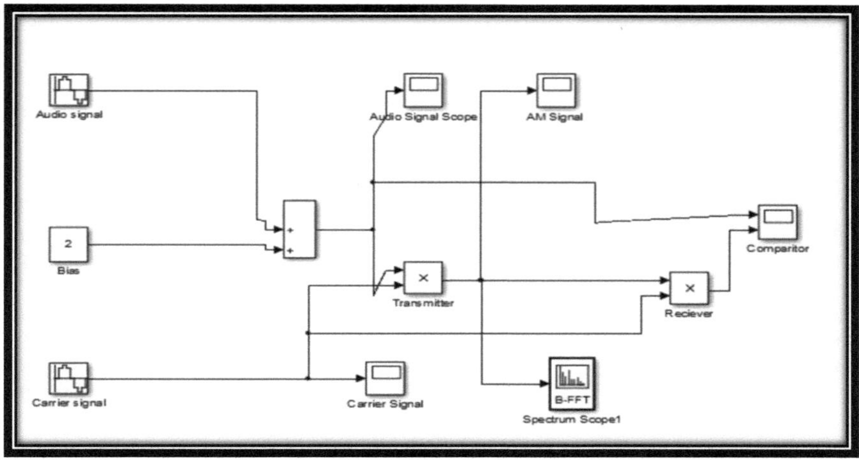

Figure 10 Simulink Model of the Receiver for Detecting the Input Audio Signal.

The following Figures show the observations of the input and output audio signal also show the output of carrier signal for the receiver

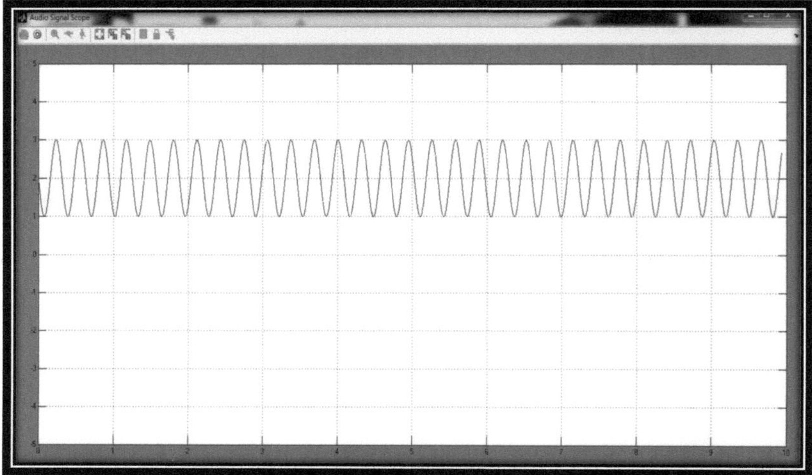

Figure 11 Audio Signal scope

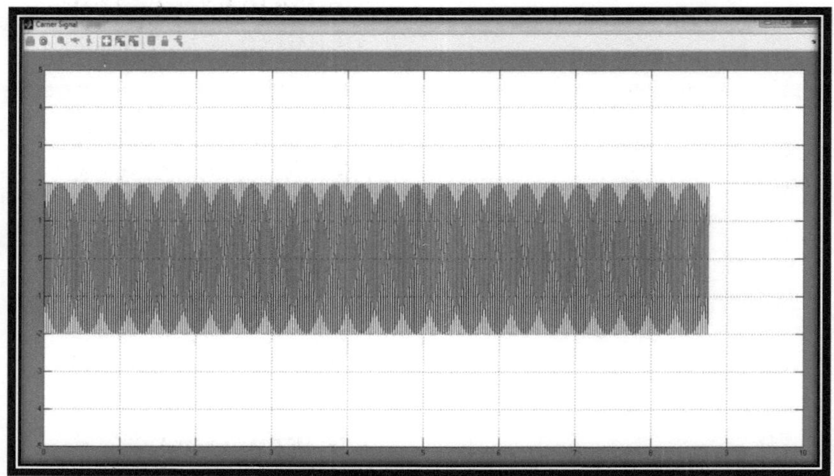

Figure 12 Carrier Signal Scope.

Figure 12 obtained by the scope connected to the carrier signal generator. The signal has amplitude of 2 and frequency of 100Hz.

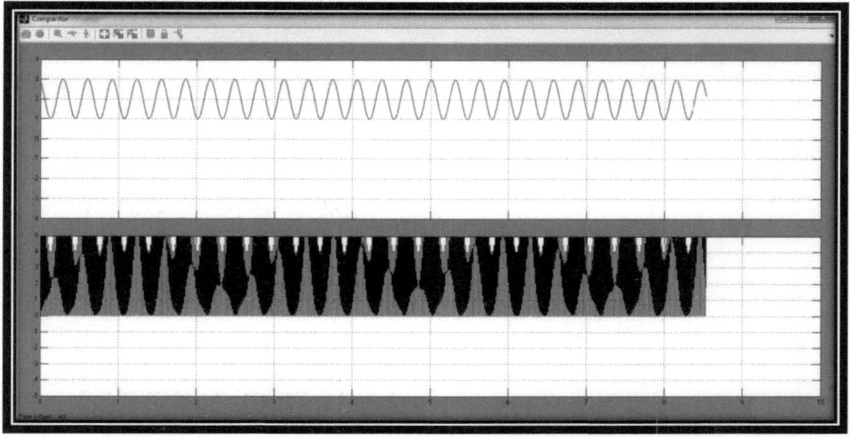

Figure 13 waveform of modulated signal.

Figure 13 obtained by the scope connected to the product block. The graph shows the modulated signal which is after the carrier signal, the audio signal, and the biased signal has all been mixed.

6. Add an AWGN block to the system and observe the receiving signal quality by setting Different SNR values

By adding the (AWGN) in the block to the system itself, the block has been change a bit as it can be seen in the figure 14 down.

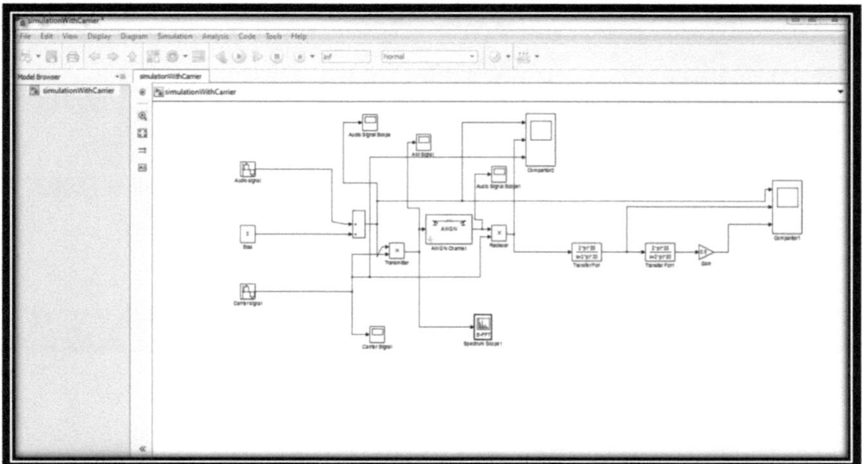

Figure 14 Add an AWGN block to the system.

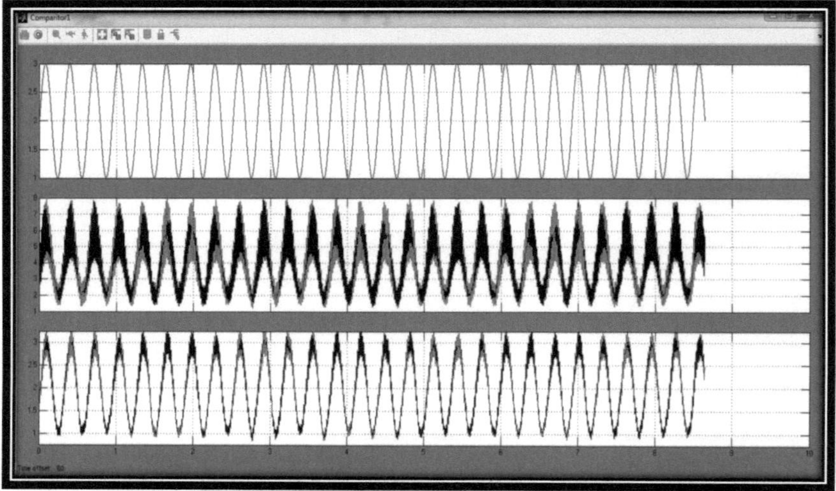

Figure 15 when SNR=0

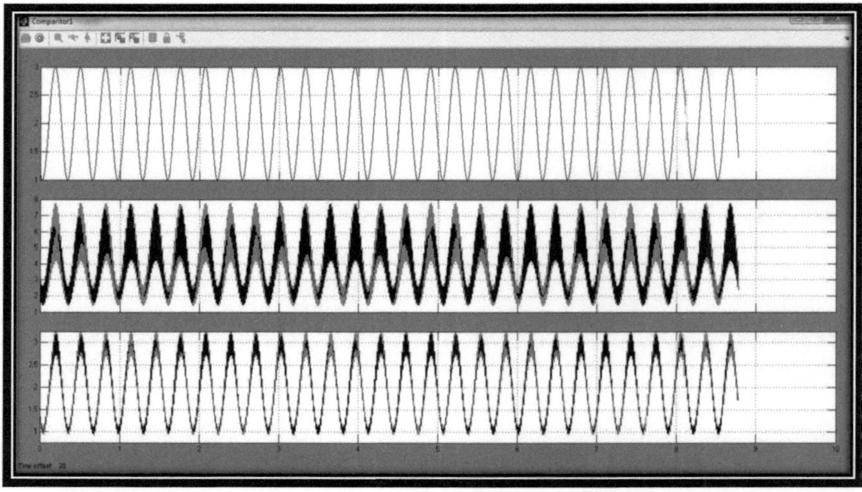

Figure 16 when SNR=20

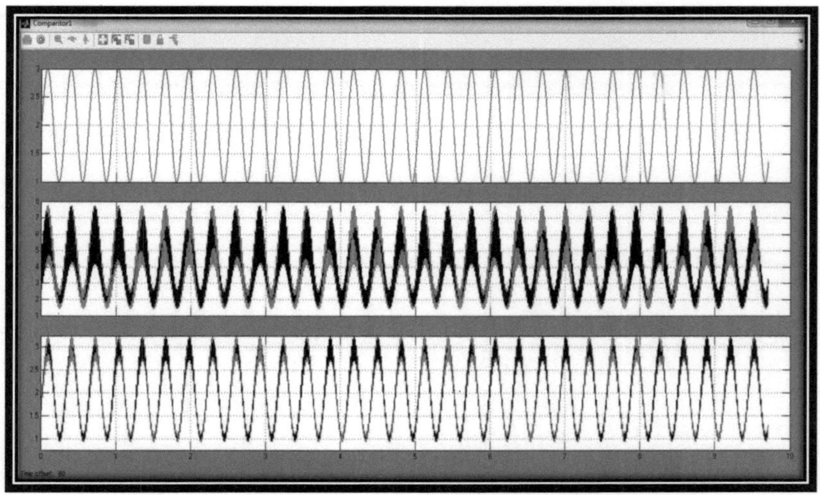

Figure 17 when SNR=60

As it can be seen from figures 15, 16 and 17 that when the user increases the SNR values from (AWGN) block, it obviously shows that receiving signal quality become more clear and clear.

7. Modify the system to communicate Double Sideband- Suppressed Carrier (DSB-SC) and Single Sideband- Suppressed Carrier (SSB-SC) signals respectively by adding appropriate filters

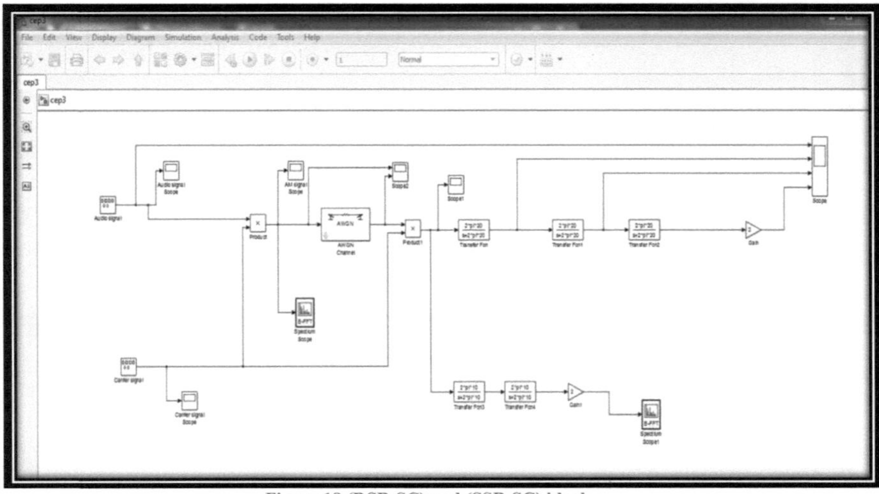

Figure 18 (DSB-SC) and (SSB-SC) block.

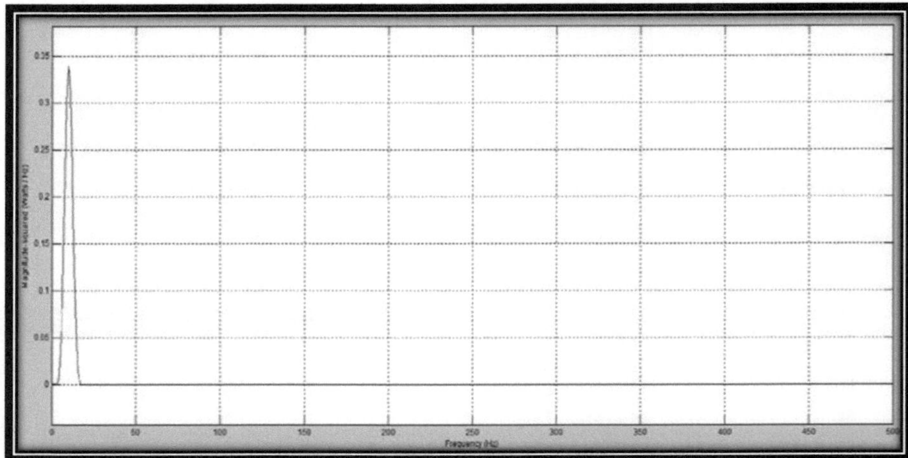

Figure 19 (SSB-SC) output frequency

In figure 19 shows that Single-sideband modulation avoids the bandwidth doubling, and the power wasted on a carrier, at the cost of increased device complexity and more difficult tuning at the receive.

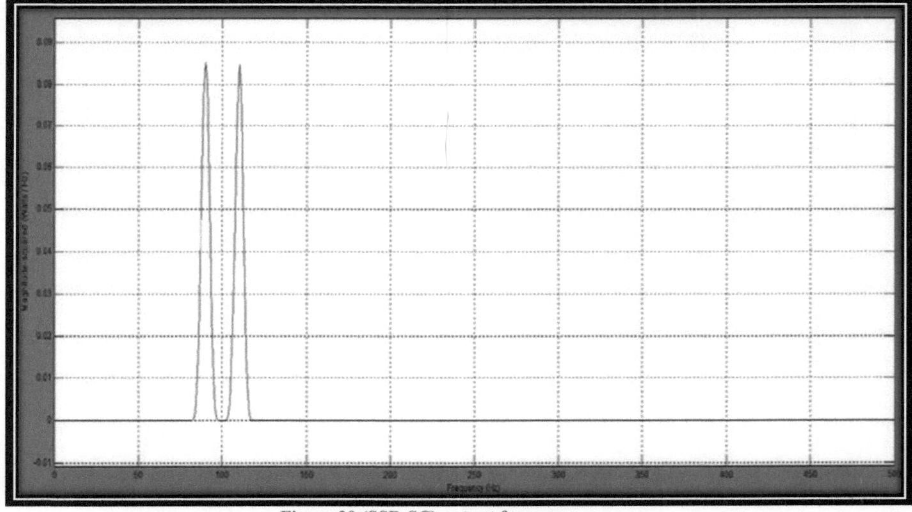

Figure 20 (SSB-SC) output frequency

As we can observe from the graph above there is no carrier frequency at 100 frequencies because in double sideband-suppressed (DSB-SC) modulation the transmitted wave consists of only the upper and lower sidebands. Transmitted power is saved through the suppression of the carrier wave.

8. Discuss whether it is better to use cascaded single order filters or higher order filter

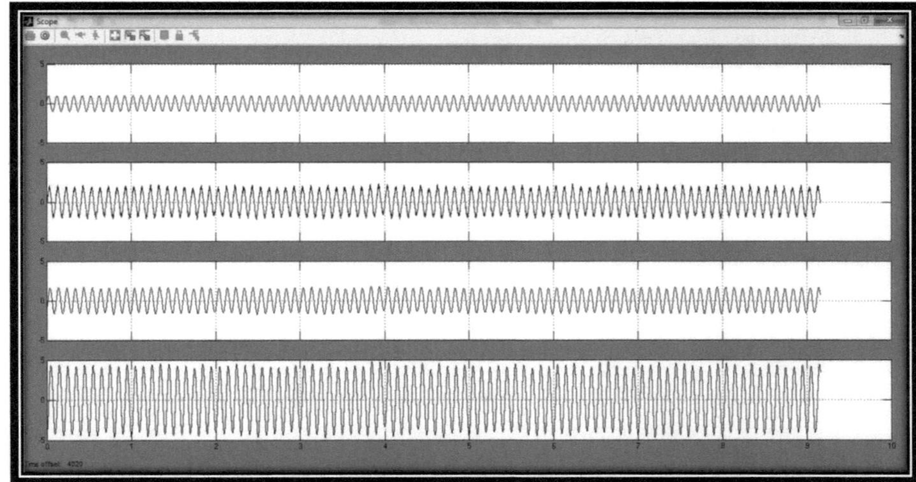

Figure 21 output waveforms with filters

As it has been seen from the graph above that when the audio signal with the first filter is looks like there some noisy showing out, same thing with the second filter bit still showing a bit clear noisy but in the last filter which it connects to the gain which it carries (gain of 3) the audio signal becomes more clear so, it's obviously seen that the higher order filter is better to use and more clear.

9. Compare the results obtained from this exercise with the theoretical results.

By comparing the results obtaining from this exercise with the theoretical results we found the modulated signal consist of carrier signal, upper sideband and lower sideband signals, and the ideal condition for amplitude modulation (AM) is when m=1 also means Am=Ac; this means the results are the same with the theoretical part.

Discussion

To get the desired output in this experiment, certain parameters need to be changed. The right parameters can be found after intense research and software testing are made.

A sine wave with 10 Hz frequency and 1V amplitude was used to represent the input audio signal. The carrier signal was assigned a value of 100 Hz frequency and 1V for the amplitude. The spectrum scope was setup correctly with the required detailed attributes to get the best visual result. These setup changes were discussed in the previous sections.

Figure 7 shows the power spectrum after keying in the parameters. The modulation index which is mainly controlled by the DC bias voltage and the amplitude is also clearly shown.

As per required in the question sheet, some parameters should be changed. Those parameters include the audio signal and the carrier signal specifications. Based on the mathematical calculations, the goal was set to achieve a modulation index of 0.4 and 0.8.

In question 5, a Simulink model of the receiver is required to be built for detecting the input audio signal and capturing the input and output waveforms. Figure 11 shows the result waveforms of the input audio signal that is assigned a frequency of 10Hz and amplitude of 1V. The carrier signal generator output is connected to the scope shown above. Here, the signal has an assigned amplitude of 2 and a frequency of 100Hz. Figure 11 shows the modulated signal which is effectively a mixture of the carrier signal, audio signal and biased signal after being well processed based on AM modulation technique.

In the question number 6, an AWGN block was added to the system model and different SNR values were used to evaluate the quality of the receiving signal. It can be seen from figure 17 that the quality of the receiving signal gets higher when the user increases the SNR values from (AWGN) block.

In question 7, the double-sideband suppressed-suppressed carrier (DSB-SC) modulation did not allow the wave carrier to be transmitted, unlike the AM modulation; thus, a great percentage of its power was distributed between the sidebands. This implies an increase of the cover in DSB-SC, compared to AM, for the same power used. DSB-SC transmission is a special case of Double-sideband reduced carrier transmission. In question 8, it was found that it is better to use cascaded higher order filters because great noise reduction and clearer audio signals were obtained as shown in figure 21.

Conclusion

In short, modulation is one of the techniques used to convey a message signal. These techniques were tested and recorded in this report. Modulation of a sine waveform is used to transform baseband message signal into pass band signal as done in this report. A low frequency audio signal is joined with a high frequency carrier signal to get a modulated signal.

We had good experience and learnt a lot about how to operate the spectrum analyzer on MATLAB and Simulink to generate and view different waveforms.

We conclude that the DSBSC is a better modulating technique because no power gets lost in the carrier side. It also noticed that each modulating frequency component produces its own upper and lower side frequencies around the carrier frequency, all the upper side frequencies are grouped together and referred to as the upper sideband (USB) and all the lower side frequencies form the lower sideband (LSB).

References

Electronics Post. (2015). Block Diagram of Communication System with Detailed Explanation - Electronics Post. [online] Available at: http://electronicspost.com/block-diagram-of-communication-system-with-detailed-explanation/ [Accessed 31 May 2016].

Hanzo, L., Webb, W. and Keller, T. (2000). Single- and multi-carrier quadrature amplitude modulation. Chichester [England]: John Wiley & Sons.

MATLAB & SIMULINK. (2010). [Natick, Mass.]: Mathworks.

Webb, W. and Hanzo, L. (1994). Modern quadrature amplitude modulation. London: Pentech.

YOUR KNOWLEDGE HAS VALUE

- We will publish your bachelor's and
  master's thesis, essays and papers

- Your own eBook and book -
  sold worldwide in all relevant shops

- Earn money with each sale

Upload your text at www.GRIN.com
and publish for free